Safari Park

by Stuart J. Murphy • illustrated by Steve Björkman

HarperCollins *Publishers*

To Janet Ginsburg—whose mind is like a safari of wild ideas
—S.J.M.

For Matthew and Timothy Gross with love
—S.B.

The publisher and author would like to thank teachers Patricia Chase, Phyllis Goldman,
and Patrick Hopfensperger for their help in making the math in MathStart just right for kids.

HarperCollins®, 🏛 ®, and MathStart® are registered trademarks of HarperCollins Publishers.
For more information about the MathStart series, write to HarperCollins Children's Books,
1350 Avenue of the Americas, New York, NY 10019, or visit our website at www.mathstartbooks.com.

Bugs incorporated in the MathStart series design were painted by Jon Buller.

Safari Park
Text copyright © 2002 by Stuart J. Murphy
Illustrations copyright © 2002 by Steve Björkman
Printed in the U.S.A. All rights reserved.

Library of Congress Cataloging-in-Publication Data
Murphy, Stuart J.
    Safari Park / by Stuart J. Murphy ; illustrated by Steve Björkman
        p.   cm.
    "Finding unknowns, level 3."
    ISBN 0-06-028914-7 — ISBN 0-06-028915-5 (lib. bdg.) — ISBN 0-06-446245-5 (pbk.)
    1. Equations—Numerical solutions—Juvenile literature.   [1. Equations.]   I. Björkman, Steve, ill.   II. Title.
QA218.M87 2002   00-063201
512.9'4—dc21   CIP
                                                                                                              AC

Typography by Elynn Cohen   1 2 3 4 5 6 7 8 9 10   ❖   First Edition

All summer long, everyone was talking about the new Safari Park. Grandpa promised to take all his grandchildren on opening day.

The supermarket was giving away free tickets for rides and games and food. The ride Paul wanted to try most was the Terrible Tarantula.

Opening day was finally here!

Grandpa had collected 100 tickets. He gave 20 to each of his 5 grandchildren, and they were off to Safari Park.

When they got to the entrance, Paul felt in his pockets. He felt in his backpack. "Oh no," he said.

"Typical," said his sister, Abby.

"It looks like Paul lost his tickets," said Grandpa. "You'll each have to take him on 1 ride."

JUNGLE KINGS

4 TICKETS EACH

WILDERNESS CARS
TREETOP COASTER
RIVER RAFT
TIGER WHEEL

"Okay!" said Grandpa. "Before you spend any of your tickets, you'd better decide what rides you want to go on. That way you'll be sure you don't run out of tickets. Who knows which rides they want to go on?"

"I do," said Chad. "I'll go on the Wilderness Cars, and Paul can come with me. And I want to go on the River Raft, too. For Rhino Rides, I want to try the Elephant Twirl, the Bat Tunnel, and the Rope Swing. Then how many Monkey Games can I play?"

"Monkey Games are 1 ticket each," said Grandpa. "You have 2 tickets left over, so you can play 2."

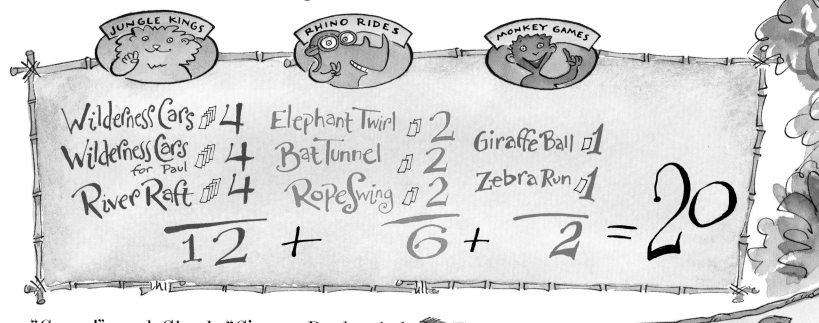

JUNGLE KINGS RHINO RIDES MONKEY GAMES

Wilderness Cars 4
Wilderness Cars for Paul 4
River Raft 4

Elephant Twirl 2
Bat Tunnel 2
Rope Swing 2

Giraffe Ball 1
Zebra Run 1

12 + 6 + 2 = 20

"Great!" said Chad. "C'mon, Paul, while everybody else is still deciding, let's go on the Wilderness Cars."

"Okay," said Paul. "But somebody's going to take me on the Tarantula, right?"

Pretty soon Chad and Paul were in their car.

ZOOM!

CARS

When they stumbled out of their car, they ran into Alicia.
"What rides are you going on?" asked Chad.

"All the Rhino Rides," Alicia said. "And Paul can come on the Elephant Twirl with me. Then I'm going to try the Rock Toss and the Snake Charmer."

"So how many Jungle Kings are you going to do?" asked Paul.

JUNGLE KINGS

RHINO RIDES

MONKEY GAMES

?
___
?

Hippo-Drome ☑ 2
Bat Tunnel ☑ 2
Rope Swing ☑ 2
Elephant Twirl ☑ 2
Elephant Twirl ☑ 2
for Paul
___
+ 10

Rock Toss ☑ 1
Snake Charmer ☑ 1
___
+ 2 = 20

"Two," answered Alicia. "River Raft and Tiger Wheel."

JUNGLE KINGS

RHINO RIDES

MONKEY GAMES

River Raft 4
Tiger Wheel 4

Hippo-Drome 2
Bat Tunnel 2
Rope Swing 2
Elephant Twirl 2
Elephant Twirl for Paul 2

Rock Toss 1
Snake Charmer 1

$8 + 10 + 2 = 20$

"I got tickets for the Elephant Twirl," said Chad. "Let's go together."

"Okay," said Paul. "But is anybody saving me tickets for the Tarantula?"

"Hurry up!" said Alicia.

Before long, Chad, Alicia, and Paul were spinning around on the Elephant Twirl.

WHEEEE!

ELEPHANT TWIRL

15

Meanwhile, Patrick was still deciding.

"I guess I'll go on the Wilderness Cars, the River Raft, and the Tiger Wheel," said Patrick. "And I'll do the Bat Tunnel and the Rope Swing and the Hippo-Drome. And I want cotton candy and a soda."

"Don't forget Paul," Grandpa said.

"Oh, yeah. He can come on the River Raft," said Patrick.

Wilderness Cars 4

River Raft 4

River Raft for Paul 4

Tiger Wheel 4

Bat Tunnel 2

Rope Swing 2

Hippo-Drome 2

Cotton Candy 1

Soda 1

16 + 6 + 2 = 24

"Wait a minute," said Grandpa. "That's 4 too many.

You only have 20 tickets."

"Oh. I guess I'll give up the Tiger Wheel," said Patrick.

JUNGLE KINGS RHINO RIDES TIGER TREATS

Wilderness Cars ⬜ 4   Bat Tunnel ⬜ 2   Cotton Candy ⬜ 1
River Raft ⬜ 4   Rope Swing ⬜ 2   Soda ⬜ 1
River Raft ⬜ 4   Hippo-Drome ⬜ 2
for Paul

$$12 + 6 + 2 = 20$$

Paul, Chad, and Alicia came
running up. "Are we going on the
Tarantula, Patrick?" asked Paul.

"No, it costs too many tickets," said
Patrick. "Let's go on the River Raft
instead."

Before they knew it, Chad, Alicia,
Patrick, and Paul were whooshing
down the River Raft.
SPLASH!

While they were drying off, they saw Abby and Grandpa.
"I got the best tickets!" Abby announced. "I got the Tiger
Wheel and the Wilderness Cars and all the Rhino Rides!
And I saved 1 ticket for a pretzel, too."
"Aren't you going to play any Monkey Games?" asked
Alicia.

Tiger Wheel 🎟 4
Wilderness Cars 🎟 4
──────────
8

Hippo-Drome 🎟 2
Elephant Twirl 🎟 2
Bat Tunnel 🎟 2
Rope Swing 🎟 2
──────────
+   8

?

+ ?

Pretzel 🎟 1

+ 1 = 20

21

"I want to do the Giraffe Ball and the Snake Charmer," said Abby. "Paul can play the Rock Toss if he wants."

JUNGLE KINGS

RHINO RIDES

MONKEY GAMES

TIGER TREATS

Tiger Wheel 4
Wilderness Cars 4
___
8 +

Hippo-Drome 2
Elephant Twirl 2
Bat Tunnel 2
Rope Swing 2
___
8 +

Giraffe Ball 1
Snake Charmer 1
Rock Toss 1
for Paul
___
3 +

Pretzel 1
___
1 = 20

All the cousins ran over to try the Monkey Games.

Soon it was Paul's turn to try the Rock Toss. *What a boring game,* he thought. *Who wants to throw a dumb rock around when you could be riding the Tarantula?*
He tossed his rock without even aiming.
"Congratulations!" said the Rock Toss lady. "You just won 18 tickets! And you get another toss!"

This time Paul aimed carefully. He held his breath. And he
threw the rock.

CLANG!

"You win!" said the lady. "You're a lucky guy!" And she gave
Paul 18 more tickets.

Paul knew *exactly* what he was going to do with the tickets.

"It's really big," said Chad. "But I'll come."

"Me too!" said Alicia.

"It won't make me sick, will it?" asked Patrick.

"No way," said Paul. "What about you, Abby?"

"Uh, I'm not sure," said Abby.

"Come on!" said Paul. "You're not scared, are you?"

"Scared?" said Abby. "I'm not scared of anything my little brother can do!"

"Yiiiikkkkessss!" screeched Chad.

"Eeeeeeeeeeekkkkk!" hollered Alicia.

"Ooooooooohhhhhhh!" moaned Patrick.

"Hellllllllppppp!" shouted Abby.

"Yeeeeeeaaaaahhhhhhh!" yelled Paul.

"Grrrreeeeeaaatttt!" agreed Grandpa.

31

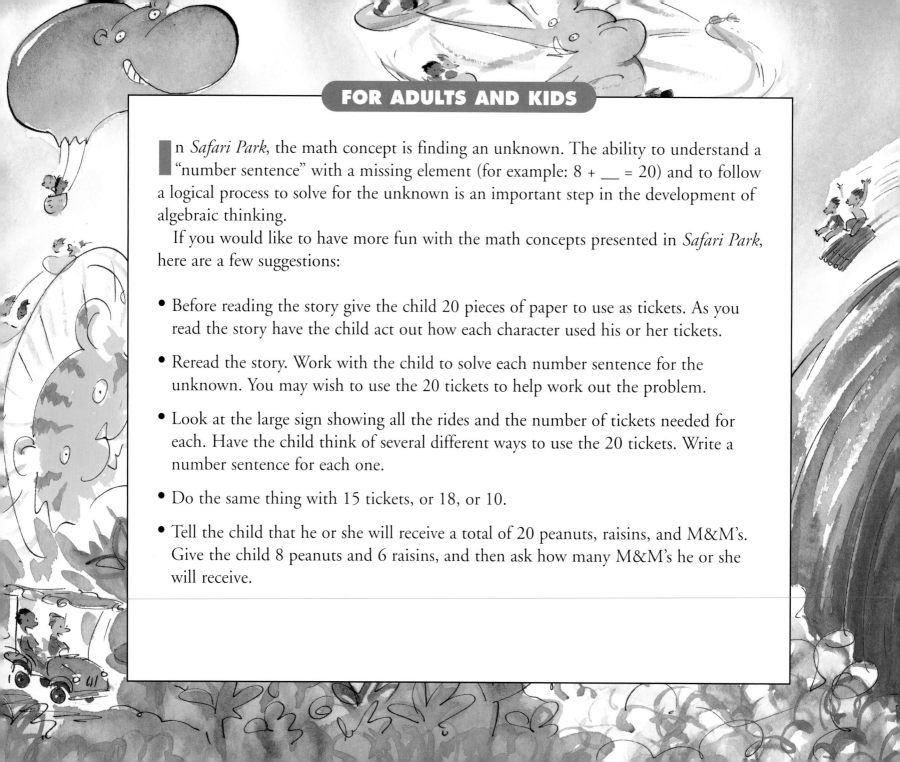

FOR ADULTS AND KIDS

In *Safari Park*, the math concept is finding an unknown. The ability to understand a "number sentence" with a missing element (for example: 8 + __ = 20) and to follow a logical process to solve for the unknown is an important step in the development of algebraic thinking.

If you would like to have more fun with the math concepts presented in *Safari Park*, here are a few suggestions:

- Before reading the story give the child 20 pieces of paper to use as tickets. As you read the story have the child act out how each character used his or her tickets.

- Reread the story. Work with the child to solve each number sentence for the unknown. You may wish to use the 20 tickets to help work out the problem.

- Look at the large sign showing all the rides and the number of tickets needed for each. Have the child think of several different ways to use the 20 tickets. Write a number sentence for each one.

- Do the same thing with 15 tickets, or 18, or 10.

- Tell the child that he or she will receive a total of 20 peanuts, raisins, and M&M's. Give the child 8 peanuts and 6 raisins, and then ask how many M&M's he or she will receive.

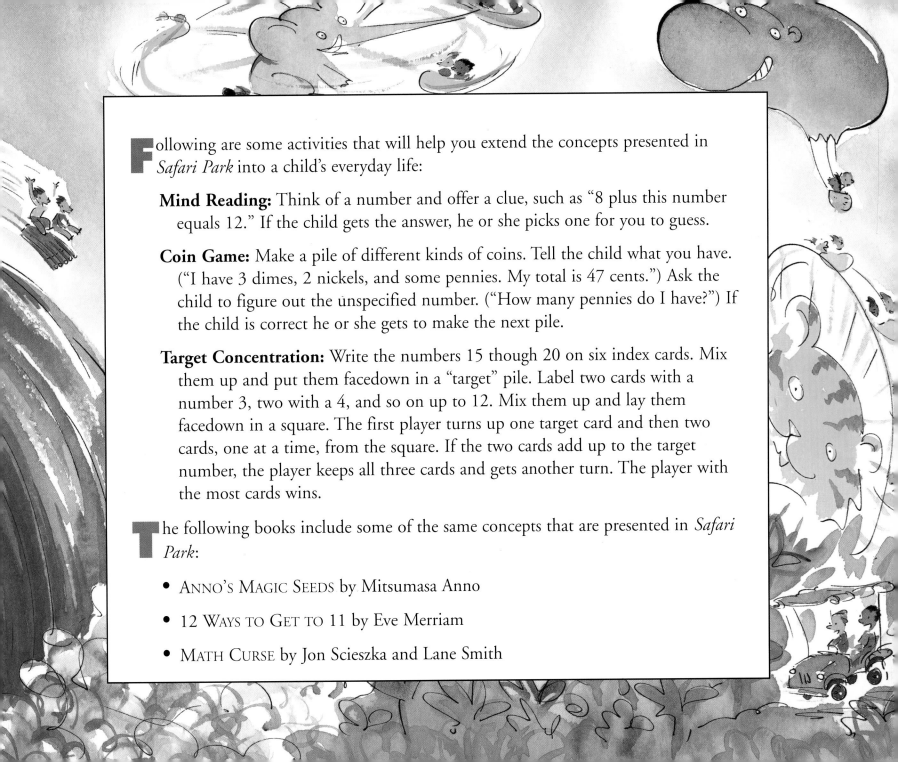

Following are some activities that will help you extend the concepts presented in *Safari Park* into a child's everyday life:

Mind Reading: Think of a number and offer a clue, such as "8 plus this number equals 12." If the child gets the answer, he or she picks one for you to guess.

Coin Game: Make a pile of different kinds of coins. Tell the child what you have. ("I have 3 dimes, 2 nickels, and some pennies. My total is 47 cents.") Ask the child to figure out the unspecified number. ("How many pennies do I have?") If the child is correct he or she gets to make the next pile.

Target Concentration: Write the numbers 15 though 20 on six index cards. Mix them up and put them facedown in a "target" pile. Label two cards with a number 3, two with a 4, and so on up to 12. Mix them up and lay them facedown in a square. The first player turns up one target card and then two cards, one at a time, from the square. If the two cards add up to the target number, the player keeps all three cards and gets another turn. The player with the most cards wins.

The following books include some of the same concepts that are presented in *Safari Park*:

- ANNO'S MAGIC SEEDS by Mitsumasa Anno

- 12 WAYS TO GET TO 11 by Eve Merriam

- MATH CURSE by Jon Scieszka and Lane Smith